The Pyramids of Giza

Scaled with Respect to Space and Time

Brent Crowder

First Edition

Fulton Books

Meadville, PA

Published by Fulton Books 2024

Printed in the United States of America

Contents

CHAPTER 1
Introduction

Did the designer of the Great Pyramid of Giza, more than 4,000 years ago, foresee not only the units of measure more precisely than a King Henry era booted foot in the year 1100, or the 1791 meter defined about 230 years ago? The modern-day foot was most recently standardized in 1959 as exactly 0.3048 meters, so this measure has now, for more than 60 years, been "subservient" to the meter. It appears that modernly measured Earth polar and equatorial circumferences were encoded in the Great Pyramid in terms of scaled length and time with an error of less than 0.05 percent. Then consider that using pure, unitless numbers, as in table 1 of chapter 4, the resulting numbers, 756 and 481.1, just happen to be the base length and height of the Great Pyramid in feet. Information herein provides, as evidence to some, that the designer, whoever/ whenever he/that was, of the shapes and sizes of the Giza Pyramids, knew in advance of the modern-day measurement units, such as feet, meters, miles, degrees, and seconds, thousands of years before such units were standardized, quantified, or even defined. While significant mathematical relations are provided to quantify and verify, for math enthusiasts, the graphic relations, nonmath-inclined readers are encouraged to view the graphics, compare actual versus calculated results, watch the linked videos (especially note 3), and to cut, fold, and tape the figures at the end and linked (per note 10) to form three scaled pyramids.

The Earth and moon dimensions (of figure 1) can be related to a 3, 4, 5 right triangle, the smallest and best-known Pythagorean triple[1] and the Earth-moon triangle, forms two side-by-side Kepler triangles,[2] which are expressed in terms of the Greek letter, φ (also referred to as phi), known as the golden mean or golden ratio. These fundamental numbers are all inherent in the geometry of the Great Pyramid, Earth, and moon dimensions of figure 1.

Furthermore, it will be shown that the Great Pyramid incorporates a very close approximation of the more commonly recognized number, pi (or π) as the ratio of the Great Pyramid's perimeter to twice its height, approximating pi to within less than 0.05 percent error. Again noting the numbers 5, 4, 3, as the Pythagorean triple, also consider the significance of 4, 3, 2, the numerical sequence of the 43,200 cycles (occurring over 86,400 seconds) of a one-meter polar pendulum (see page 13) in a day, as well as 43,200 being the scaling factor (consistent with note 7, hour 1) of the perimeter of the Great Pyramid's foundational base to the circumference of Earth. An interesting calculation (even for nonmaths): $432 \times 432 = 186{,}624$, which is very close to the speed of light in miles per second; and note that in 1884, composer Giuseppe Verdi strived to deem 432 hertz as the standard frequency for musical note A.

While the content below provides mathematical relations to the geometry of the Giza pyramids, there is a means of drafting and constructing scale models of the three main Giza pyramids without a ruler, using nothing more than paper, tape, scissors, a straight edge, and a compass, or, if so desired, the resulting last two pages (in chapter 9) can be copied via the link of note 10, folded, and taped, to form scaled 3D pyramids.

There is no "Conclusions" chapter herein because it seems conclusions might most appropriately be individually formulated by each reader.

CHAPTER 2

Scaling the Giza Pyramids: Menkaure,

Khafre, and Khufu (in Time and Space) from the Vesica Piscis, to the Fibonacci Spiral,

the Lucas Spiral, and to Earth and the Moon

The Great Pyramid, Khufu, being the largest and having the most sophisticated external and internal characteristics, is probably the most studied pyramid. The following provides an amazing perspective as if the moon were set atop the earth, scaling the Great Pyramid in relation to earth and the moon:

Figure 1

Earth–moon–Khufu

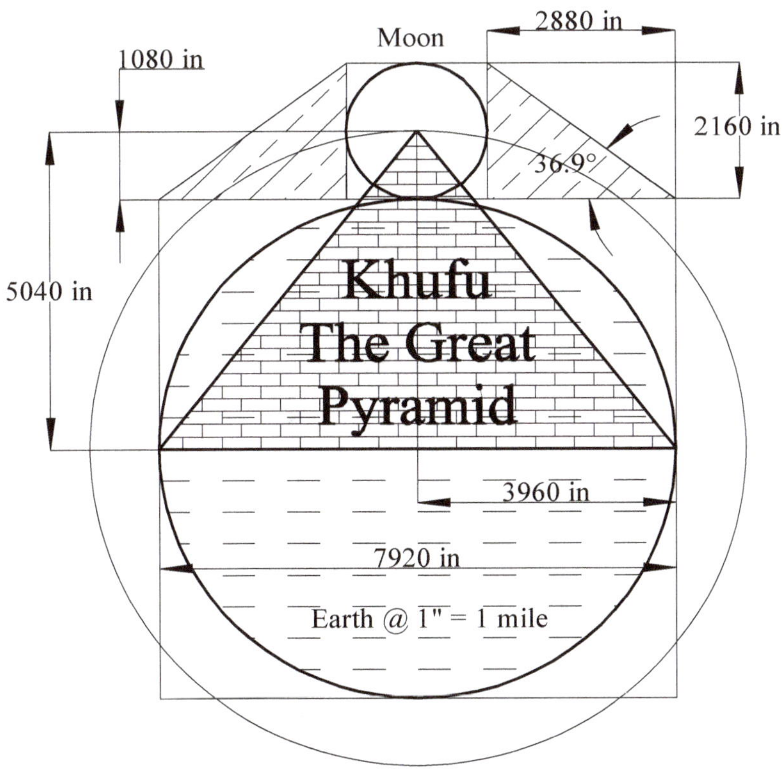

The perimeter of the square, 4 × 7,920 = 31,680 and 2 × π × 5,040 = 31,667 (squaring the circle) appear as evidence of knowledge of the relation of Khufu to Earth as well as a very close approximation of π. The triangles adjacent to the moon have the quotient, 2,160/2,880 = (3 × 720)/(4 × 720) = 0.75, the tangent of 36.9 degrees, the angle of a 3 (height) × 4 (base) × 5 (hypotenuse) triangle.

Figure 6 shows the pyramid, Khafre has 1.5 (half its base), 2 (height), 2.5 (apothem) proportions, which, scaled by a factor of 2 are 3, 4, 5 proportions. The pyramid height in the drawing above is approximately the Earth mean radius (3,960 (3,958.76 rounded up to 3,960)) miles multiplied by the square root of the golden ratio (phi or Φ): 3,960 × phi^0.5 = 5,037.2, very close to the above shown 5,040 in., which exceeds the phi-based value by a tiny 0.056%. The Great Pyramid, viewed in profile from north, south, east or west, scaled down to its profile view, for proportioning is:

Figure 2

The Scaling and Expression of the Great Pyramid(Khufu) of Giza in Terms of the Golden Proportion, Φ

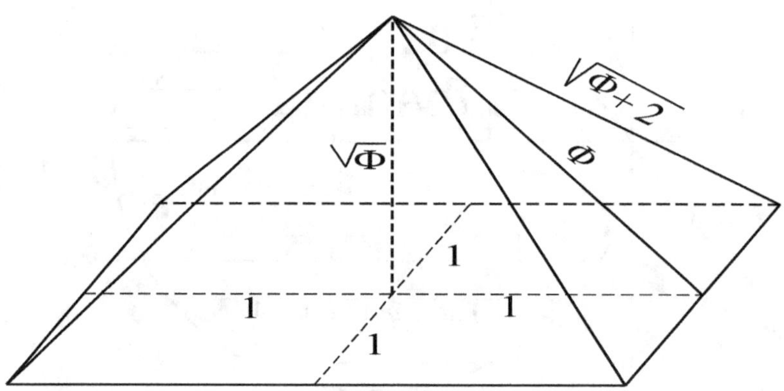

For two- and three-dimensional scaling purposes, via only a compass and straight edge, the proportion of the figure above will be created incorporating the properties of the golden ratio, phi, also expressed as the Greek letter, φ, having the unique property which satisfies the positive value for X, such that $1 + X = X^2$, or $X^2 - X - 1 = 0$, where, via the quadratic formula:

X = (-(-1) +/- sqrt ((-1)^2 – 4 × 1 × (-1))/(2 × (1) = (1 + sqrt(1 + 4))/2 = (1 +/- sqrt(5))/2. The positive value of X is X = phi = 1.618.

The following procedure describes and shows construction of the above as a foldable cutout at a factor of 2. Note, from Pythagorean theorem: 1^2 + sqrt(phi)^2 = 1 + phi = phi^2. The square root of phi^2 = phi.

The half-base distance is 756 ft/2 = 378 ft. Note, for later consideration regarding chapters 4 and 5, that 378 ft × the square root of phi = 480.82 ft. This compares closely (to within 0.06%) of 481.09 from table 1 (of chapter 4) as the height of Khufu. Also, recognize from figure 5 of chapter 5 that the 7/11 height-to-base ratio of Khufu corresponds to its half base (11/2 = 5.5) × sqrt(phi) = 5.5 × 1.272 = 6.996, confirming the 7/11 height/base ratio. The apothem (distance from the center of the 756 ft base to the top of the pyramid) is 378 × phi = 611.6 ft, closely agreeing with the 612 ft per the fourth note from Joseph Turbeville's "Table - 1 Notes."[7]

Table 1, herein, also gives rise to the significance of the number 432 in scaling, in size, consistent with the sixth note of Turbeville's "Table - 1 Notes." Then consider (again) the potential significance of this number with respect to time per the chapter 5 one-meter pendulum.

CHAPTER 3

3D Construction of the Great Pyramid (Khufu) of Giza Proportions Using a Compass and Straight Edge

The following is created based on: https://www.youtube.com/watch?v=rM2O4EdA1_A[3]

1. Draw a construction line.
2. Centered on the line (at x = 0), draw a reference circle, r = 1.
3. Draw two circles, r = 2, centered on the intersections with the two intersections of the construction line and the initial circle, forming a vesica piscis.
4. Draw a perpendicular construction line (y axis) through the center.
5. Draw an r = 2 circle at the center.
6. Note that $1^2 + 2^2 = 5$, so the hypotenuse of the red line is $5^{0.5}$. Draw the red arc, centered at 1 with r = $5^{0.5}$. The intersection of the arc and the line occurs at X = 1 + $5^{0.5}$, so X = 2phi.
7. Draw an r = 2 arc centered at 1+$5^{0.5}$ to intersect the r = 2 circle about the origin to bisect 2phi, which is phi.
8. Using r = 2phi, draw two arcs centered from X = +2 and -2 to define 4 as the base length and 2phi as the length along the face from the top of the pyramid to the center of its baseline.
9. Draw the two blue apothem—length lines that extend from the centers of the base lines and intersect at the height of the pyramid.
10. Strike the blue radius arc of length, 2phi centered on the intersection (origin) of the y axis and the construction line with the arc extending from the construction line up to the y axis.
11. The apothem length is phi times the half base length. Draw the edge length as the hypotenuse formed by a right triangle resulting from the two legs, 2 and 2phi. Thus, the edge length is 2 × (2 + phi)$^{0.5}$ as determined in figure 3.

12. Draw a circle with the radius of the above-described hypotenuse to define the lengths of the four aboveground pyramid edges.

13. Use $r = 4$ to mark the remaining (three plus one overlapping) base cord length vertices for the pyramid green base edges.

14. To create the square base, provide an extension segment for the purpose of creating a ninety-degree angle using the extension segment endpoint and its opposing arc strike intersection to create the associated bisecting arcs.

15. Complete the base square segment lengths by appropriately striking three more $r = 4$ base edge length arcs.

16. Draw the square diagonals to locate the center of the square base.

17. Draw a circle inscribing the base via its center. This will serve as a check to ensure the circle and square vertices coincide.

Figure 3

Proportions of Khufu in terms of phi

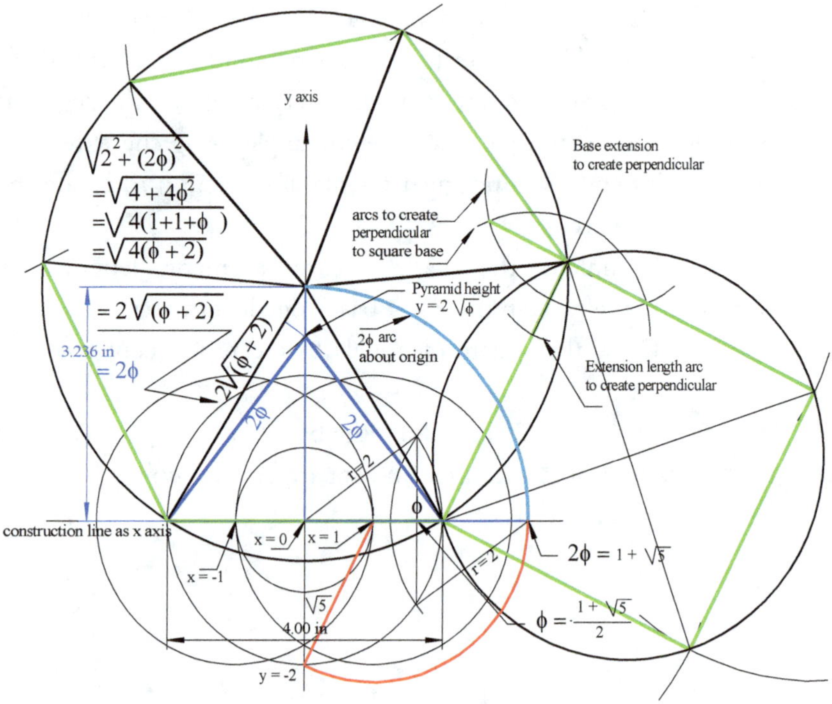

In order to scale the Great Pyramid from a one-mile-per-inch perspective, as in figure 1, as well as a 1,000 scale-up of "4" base length (figure 3, above), the following 4,000-inch base layout will serve as a basis for scaling to the Khufu dimensions.

Figure 4

Scaling of figure 3 × 1,000

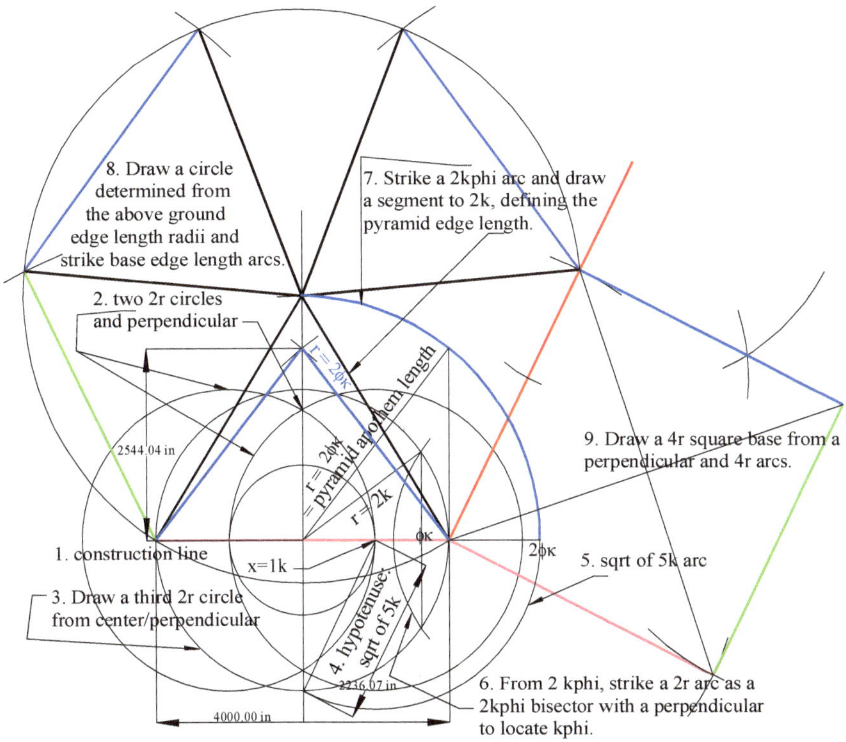

Figure 4 is consistent with the above compass–straight edge Khufu construction video scaled up by a factor of 1,000. Dimensions herein are typically shown on a scale of thousands of inches to represent miles (1" = 1 mile) and/or cubits (example: 4,400 inches = 440 cubits). Note that figure 4 scaled up by a factor of 2 would provide a Khufu-proportioned pyramid with a base = 8,000 inches. Then scaling that by 0.99 results in a scaled Khufu with a base of 8,000 × 0.99 = 7,920 inches and a height of 5,037 inches (very close to 5,040 inches) as in figure 1. Also note that the scaling of Khufu in figure 6 below by a factor of 0.72 results in a base of 11,000 inches × 0.72 = 7,920 inches and 0.72 × 7,000 = 5,040 inches as in figure 1.

Evidence (https://www.willemwitteveen.com/the-royal-cubit/)[4] suggests the Giza pyramids were constructed by measurements of Egyptian Royal Cubits. This cubit is essentially (or very close to) the arc length of a pendulum of one-meter radius, swinging through an angle of 30 degrees (15 degrees each side of the vertical position). This results in an arc length of $\pi/6$ meters = 0.5236 meters = 20.614 inches, which approximately equals 1 cubit, so cubits (or thousands of cubits) will be used as the unit of measure for proportioning the pyramids in figure 6. Also noteworthy is that the rectangular boxshaped King's Chamber within Khufu exhibits the proportions of a 3, 4, 5 triangle formed by the longest corner–corner diagonal (25 cubits), the horizontal length (20 cubits), and the end diagonals (15 cubits). Furthermore, the diagonal of the two height by length walls is the square root of 525 cubits. $525^{0.5} \times 0.5236 = 11.997$ meters, extremely close (by coincidence?) to 12 meters.

Two- and three-dimensional proportioning of the pyramids of Giza is addressed per the following excerpt from chapter 5, figure 5. See minute 2:36 from: https://www.youtube.com/watch?v=ut0FBHe4nmc[5] .

Fibonacci series: 0, 1, 1, 2, 3, 5, 8, 13, 21, 34, 55, 89, 144, 233, 377, 610, 987, 1597, 2584, 4181, 6765, 10946, 17711, 28657, 46368…

Lucas series: 1, 3, 4, 7, 11, 18, 29, 47…

CHAPTER 4

Relations: One-Meter Pendulum, Great Pyramid, Earth, Time, Fibonacci Series, and the Foot

Before proceeding to dimensions, a process known as distillation can be applied to the Fibonacci series. This process expresses each multidigit number as a single-digit sum. For example, the number 13 is expressed as 1 + 3 = 4, and the twenty-fourth Fibonacci series number after 0, 46368, as 27 (2 + 7) = 9. Thus, the distilled Fibonacci series (excluding 0) becomes a repetition of 1, 1, 2, 3, 5, 8, 4, 3, 7, 1, 8, 9, 8, 8, 7, 6, 4, 1, 5, 6, 2, 8, 1, 9, 1, 1, 2, 3, 5, 8, 4, 3, 7, 1, 8, 9, 8, 8, 7, 6, 4, 1, 5, 6, 2, 8, 1, 9 from: https://old.world-mysteries.com › jto_article2.pdf [6] .

The following table, although not explicitly from the revelation by Joseph Turbeville, is derived from his work.

Table 1

Dimensionless distillation of the Fibonacci series resulting in the base length and height of Khufu in feet

	Distilled Sum of row												Row	9's	2 x Row	2 x 9's
	1	2	3	4	5	6	7	8	9	10	11	12	Sum	sum	Sum	sum
0	1	1	2	3	5	8	4	3	7	1	8	9	52	9	104	18
0	2	2	4	6	1	7	8	6	5	2	7	9	59	9	118	18
0	3	3	6	9	6	6	3	9	3	3	6	9	66	27	132	54
0	4	4	8	3	2	5	7	3	1	4	5	9	55	9	110	18
0	5	5	1	6	7	4	2	6	8	5	4	9	62	9	124	18
0	6	6	3	9	3	3	6	9	6	6	3	9	69	27	138	54
0	7	7	5	3	8	2	1	3	4	7	2	9	58	9	116	18
0	8	8	7	6	4	1	5	6	2	8	1	9	65	9	130	18
0	9	9	9	9	9	9	9	9	9	9	9	9	108	108	216	216
0	45	45	45	54	45	45	45	54	45	45	45	81	594	216	1188	432
Sum of above numbers in column																

Casting out of the 9's provides the (**square** base), in feet, of Khufu:	1188- 432 =	756
Square base area:	756 x 756 =	571,536
Khufu height in feet	571,536/1,188 =	481.091

Turbeville (from row 1, above) also discovered the following distilled Fibonacci series-based numerator/denominator method of obtaining inches in a cubit: 1/1 + 2/1 + 3/2 + 5/3 + 8/5 + 4/8 + 3/4 + 7/3 + 1/7 + 8/1 + 9/8 = 20.61785714 or 20.618, very close to the value of 20.614" obtained by the arc length of a one-meter pendulum resulting from a $\pi/6$ (30-degree) angle discussed above. Note that 440 cubits × 20.618 inches/cubit × 1 foot/12 inches = 755.99 feet as determined (or rediscovered) by Joseph Turbeville. Turbeville arrives at a Khufu base of 756 and height of 481.1 derived from his unitless tabular method (table 1), corresponding to 280/440 × 756 ft = 481.1 ft. Thus, the dimensions of Khufu include indications that the cubit, the meter, inches, and feet were all incorporated into its measurements. The 756-ft Khufu base does not include the socle, which extends, as a foundation, slightly beyond the base, resulting in a 760.9-ft base length including the socle per hour 1, minute 0 of https://www.youtube.com/watch?v=R7oyZGW99os[7] . Multiply this base perimeter by **43,200** (for reasons provided below in "Khufu Scaled in Geometry and Time") to obtain the Earth's equatorial circumference = 760.9 × 4 × 43,200 = 131,483,520 ft = 24,902 miles. This corresponds to an Earth equatorial radius of 3,963.3 miles versus 3,963.2 miles, per https://nssdc.gsfc.nasa.gov/planetary/factsheet/earthfact.html[8] . Said foundation (socle) creates a height of 482.7575 ft corresponding to a calculated Earth polar radius:

43,200 × 482.7575 ft = 20,855,152 ft = 3,949.82 miles versus 3,949.9 miles from the above-linked fact sheet. So next we gain insight into the 43,200 time and space "multiplier."

CHAPTER 5

Khufu Scaled in Geometry and Time

The period of time, T, for a simple pendulum of length, L, minutely displaced from vertical, when released, to return to its position of release, exhibits simple harmonic motion where $T = 2\pi\ (L/g)^{0.5}$ with g as the local acceleration due to gravity. $T^2 = 4\pi^2 \times (L/g)$. It should be noted that g, determined below, at the poles, needs no correction for Earth's rotation, as is required for any latitude between the poles as well as the largest radius at the equator.

$T^2 = 4\pi^2\ (L/g)$ $\quad L = \dfrac{T^2 g}{4\pi^2}$ $\quad$ Solving for g at the poles:

m_p = the mass of a pendulum $\quad R_{pole}$ = Earth polar radius $\quad R_{eq}$ = Earth equator radius

R_{pole} = 6356.752 km = 6,356,752 m $\quad R_{eq}$ = 6378.137 km = 6,378,137 m

G = the gravitational constant $= \dfrac{6.6743 \times 10^{-11}\ m^3}{kg\ sec^2}$

m_e = the mass of Earth $\quad m_e = 5.9722 \times 10^{24}$ kg

Newton's law of gravitation for force of gravity, $f = G \times \dfrac{m_p \times m_e}{R^2} = m_p \times g$ (with m_p struck out)

where $g_{pole} = \dfrac{G \times m_e}{R^2_{pole}}$ $\quad g_{pole} = \dfrac{6.6743 \times 10^{-11}\ m^3 \times 5.9722 \times 10^{24}\ kg}{kg\ sec^2 \times (6{,}356{,}752\ m)^2}$

$g_{pole} = \dfrac{9.8644\ m}{sec^2} = \dfrac{9.8644\ m\ ft}{sec^2 \times .3048\ m} = \dfrac{32.363\ ft}{sec^2}$ $\quad g_{pole}$ needs no correction for Earth rotation.

Setting T = 2 sec: $L = T^2 \dfrac{g_{pole}}{4\pi^2} = \dfrac{9.8644\ m}{\pi^2}$ = 0.99947 meter = 999.5 mm

999.5 mm is one meter - 0.5 mm

$\dfrac{24\ hr}{day} \times \dfrac{60\ min}{hr} \times \dfrac{60\ sec}{min} = \dfrac{86{,}400\ sec}{day} \times \dfrac{period}{2\ sec}$ **= 43,200 periods / day**

Thus, a one-meter (999.5 mm) pendulum at a pole would cycle 43,200 times in a day, rendering the second (86,400 sec/day) a fundamental unit of time and 43,200 as the scaling factor of the Great Pyramid foundation perimeter to the Earth's equatorial circumference and its height above its foundation relative to the polar radius. The daily cycles of the meter (a measure, seemingly undefined by man at whatever time of the Great Pyramid's construction) pendulum period corresponding to the Earth/pyramid scaling does not appear to be accidental.

In that day there will be an altar to the Lord in the midst of the land of Egypt, and a pillar to the Lord at its border. It will be a sign and a witness to the Lord of hosts in the land of Egypt. (Isaiah 19:19–20)[9]

Figure 5

Proportions of Menkaure, Khafre, and Khufu

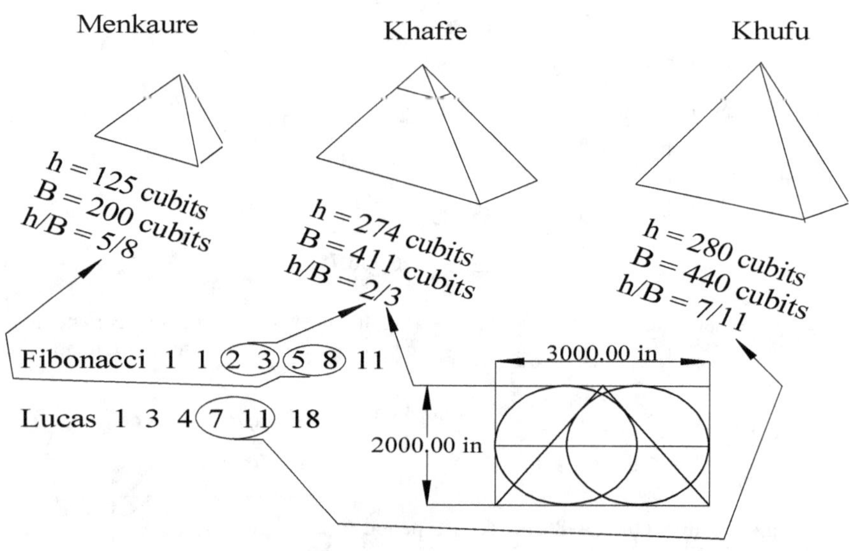

The Pyramids of Giza

The following is scaled to 1 inch = 1 mile in the spirit of dimensioning the Great Pyramid, rounded to low whole number Fibonacci/ Lucas series multiples in thousands of "inches/miles," resulting in a Khufu base of 11,000, while also being "Lucas spiraled" based on initial radii of two one-thousand-inch circles forming the vesica piscis and using the 2/3 height to base, h/B, ratio of Khafre, the smallest (green) pyramid in figure 6 below. It is important to recognize that figure 6 is for proportioning (via the Fibonacci and Lucas spirals) the shapes, but not the relative sizes of Khafre (green), Menkaure (blue), and Khufu (red) pyramids.

Figure 6

The spiraled proportions of the Giza Pyramids

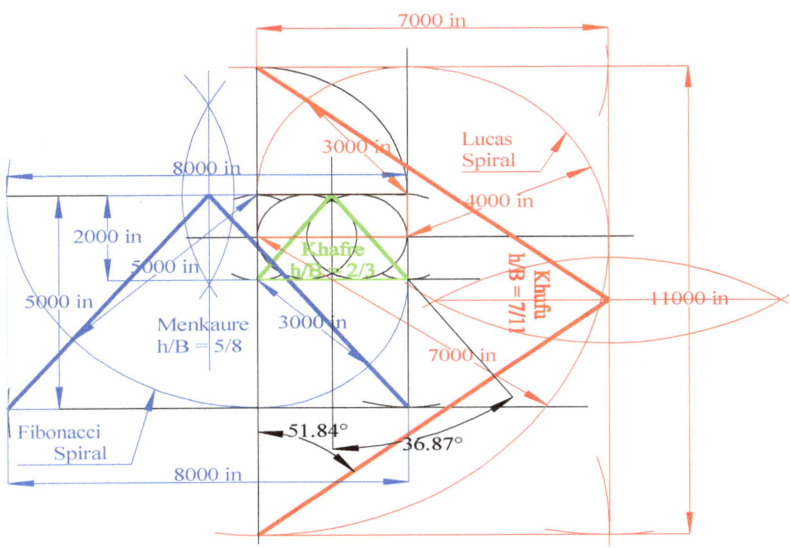

Figure 6, above, does not provide the sizes of the pyramids relative to each other. It provides only the height-to-base ratios of each pyramid.

CHAPTER 6
Scaling Menkaure

The proportions within each of the three pyramids have been established. The relative sizes of Khafre, Menkaure, and Khufu will be geometrically determined (using only a compass and a straight edge) relative to each other based on figure 5 base lengths of 200, 411, and 440 cubits. First, use figure 6 of Menkaure, with base length 8,000 inches, bisecting the base and height twice to obtain the 2,000 cubit (1 cubit = 1 inch) base and 1,250 cubit height of Menkaure.

Figure 7

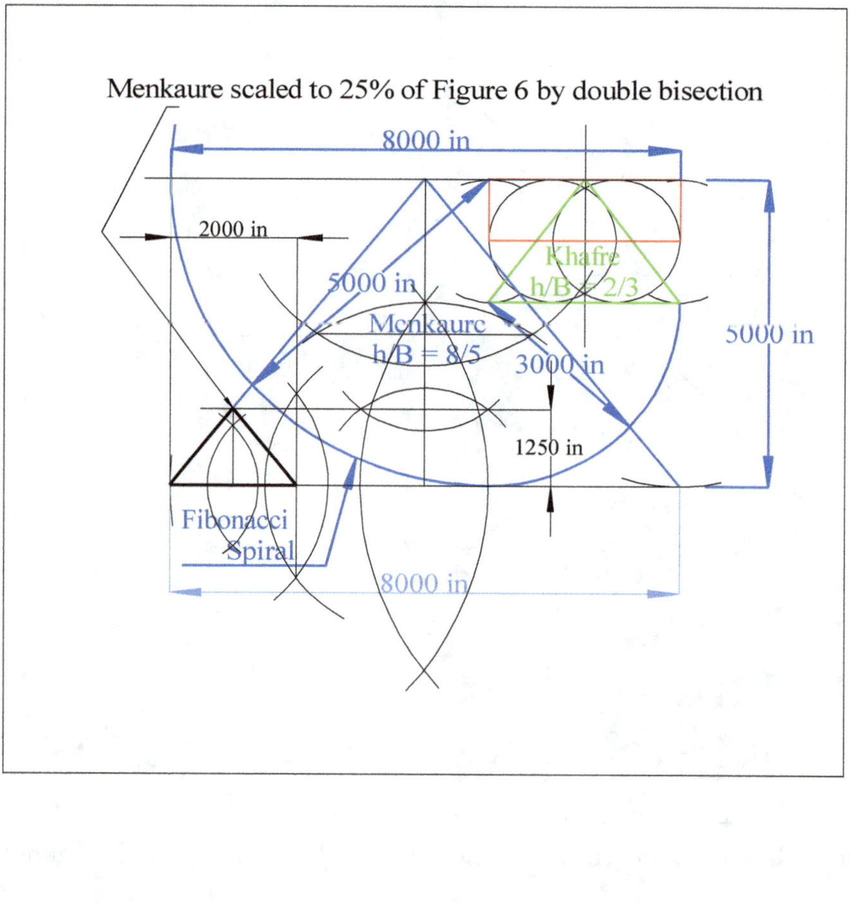

CHAPTER 7
Scaling Khafre

In order to scale Khafre, note that the base length and height measurements per figure 5 are 3 and 2 multiples of prime number 137. Figure 8, below, shows the blue line that will be used to scale up to the Khafre base length of 4,110 using the method in figure 9 to obtain 120.

Figure 8

Scaling from Menkaure to Khafre

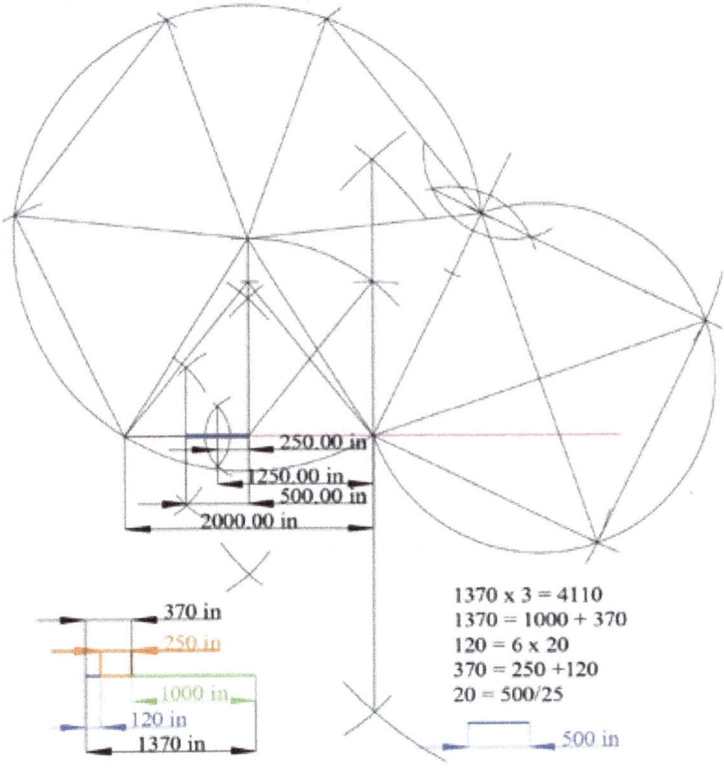

Figure 9

The scaling of Khafre to obtain 120 inches

The purpose below is to use the 500 inch segment, AB, scaled from Menkaure, above, to obtain a 120 inch segment The 120 inch segment is to be added to the 250 inch segment from Menkaure, thus obtaining a 370 inch segment, which is added to its 1000 inch segment to obtain a 1370 inch segment, which will be multiplied by 3 to obtain the base length of Khafre. 3 x 1370 in. = 4110 in. The grahic below is enlarged for illustration.

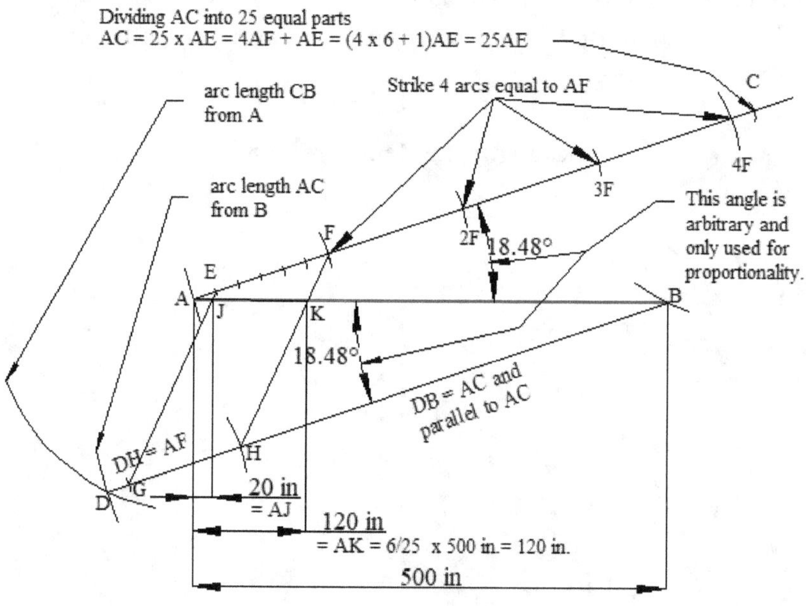

The 120-inch length, AK, from figure 9 is added to the red line (250+ 120 = 370) to get the 1,370-inch line in figure 8 and is arced three times to form the base and two times for the height of figure 10.

Figure 10 Khafre scaled to 1 inch = 0.1 cubits

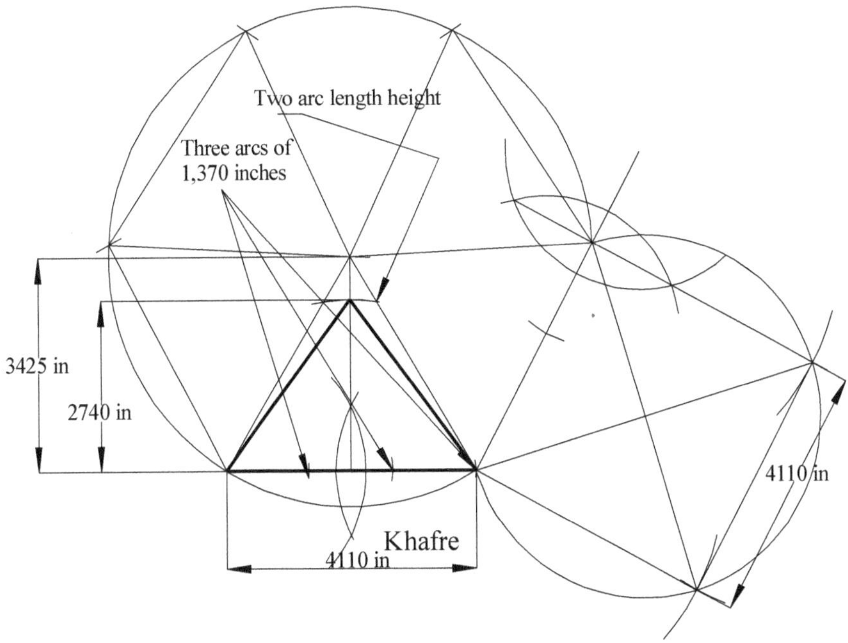

CHAPTER 8

Scaling Khufu and Comparing the Sizes of the Three Giza Pyramids

Khufu is scaled from a base dimension of 11,000 to 4,400 corresponding to 4,400/11,000 = 2/5, using the technique from figure 9.

Figure 11

Scaled base and height of Khufu to 440 cubits and 280 cubits and size comparison

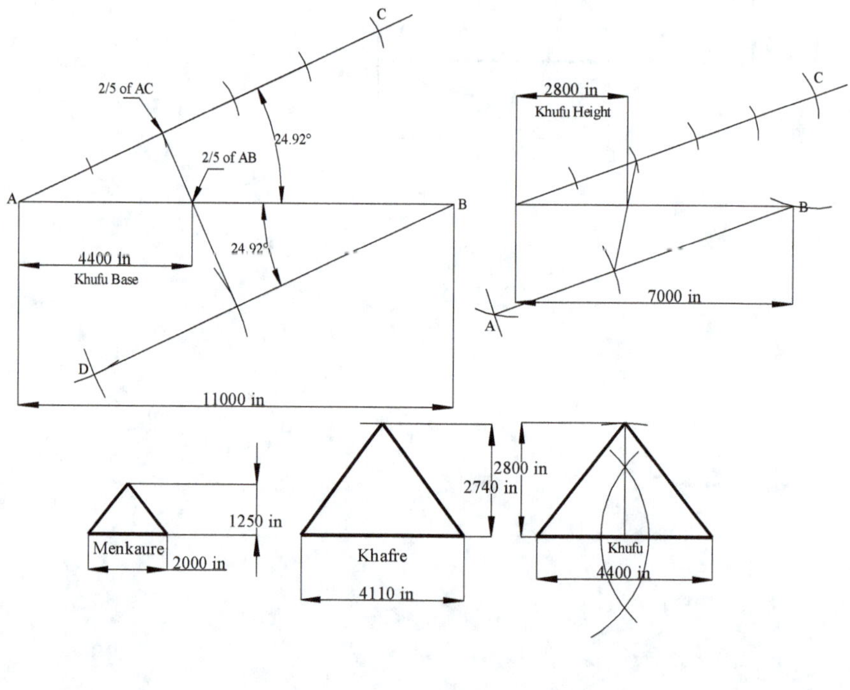

Having established the 4,400 base of Khufu, the scaled cutout pattern is depicted below in figure 12.

Figure 12

Khufu scaled to 1 inch = 0.1 cubits

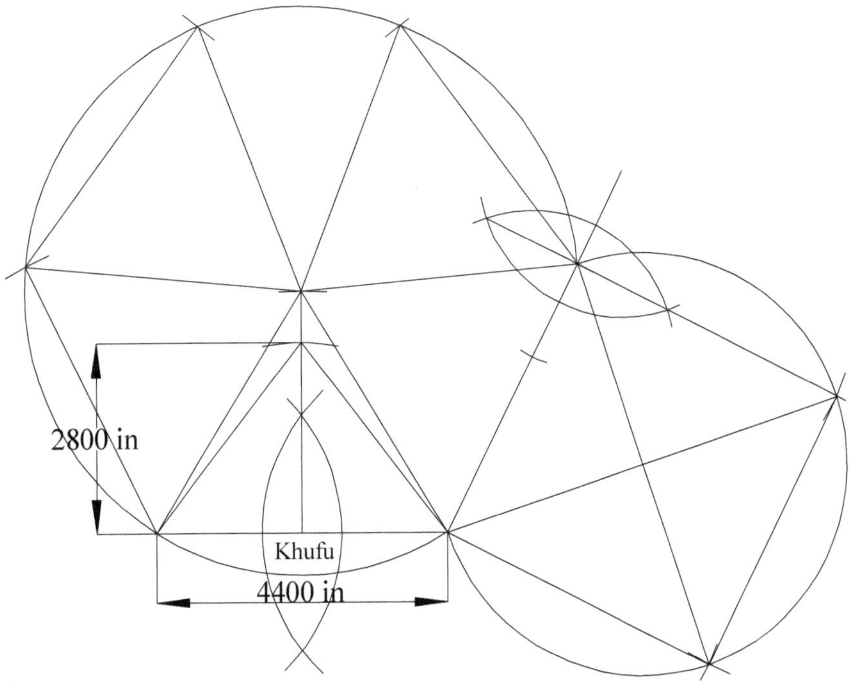

CHAPTER 9
Scaling the Three Pyramids of Giza

The last two pages, scaled to proper relative portions of the three pyramids, are available here: https://blue-claudetta-65.tiiny.site[10]

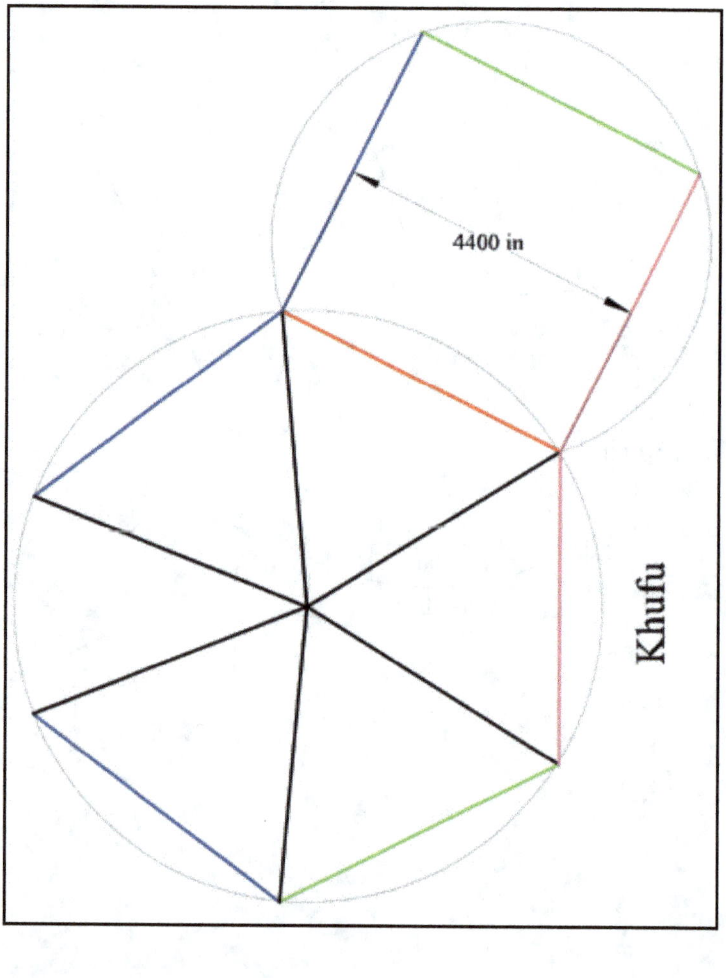

3D cutout for Khufu to same scale as Menkaure and Khafre

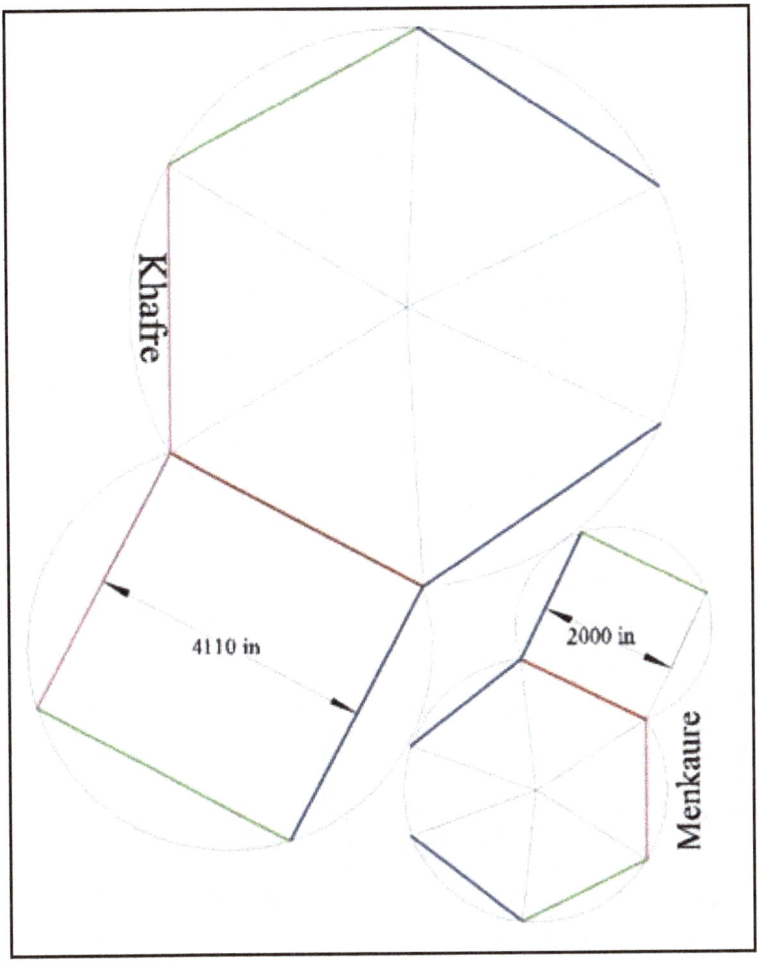

3D cutout for Khafre and Menkaure to same scale as Khufu

NOTES

1 Eric W. Weisstein, "Wolfram Mathworld, Pythagorean Triple" https://mathworld.wolfram.com/PythagoreanTriple.html

Pythagorean Triple

2 "Phi and Fibonacci in Kepler and Golden Triangles" https://www.goldennumber.net/triangles/

Golden Triangles

3 SGD Geometry Decoded, " The Great Pyramid: How to Draw in 2D & Make in 3D," YouTube video, https://www.youtube.com/watch?v=rM2O4EdA1_A

Draw the Great Pyramid

---

4 Willem Witteveen, "The Royal Cubit," https://www.willemwitteveen.com/the-royal-cubit/

Royal Cubit

5 SGD Geometry Decoded, "Giza Pyramids via the Vesica, Fibonacci, Lucas Spirals Pythagorean Music," https://www.youtube.com/watch?v=ut0FBHe4nmc

Pyramids Via Spirals

6 Joseph Turbeville, "Conjoined Values of the Royal Cubit & the Golden Ratio "Unearthed from First Twelve Terms of Fibonacci's Series" 2004

https://old.world-mysteries.com/jto_article2.pdf

Royal Cubit

7 After School, "HIDDEN MATHEMATICS – Randall Carlson – Ancient Knowledge of Space, Time & Cosmic Cycles" YouTube video, https://www.youtube.com/watch?v=R7oyZGW99os

Hidden Mathamatics

8 David R. Williams, "Earth Fact Sheet," NSSDCA, https://nssdc.gsfc.nasa.gov/planetary/factsheet/earthfact.html

Earth Fact Sheet

9 Isaiah 19:19-20 https://www.biblegateway.com/passage/?search=Isaiah+19%3A19-20&version=ESV

10 Scaled cut-outs of the three Giza pyramids: https://blue-claudetta-65.tiiny.site

Scaled cut-outs

*A cautionary note regarding the notes above and associated links/QR codes: The intent of the links and QR codes associated with the notes herein is to provide the reader with convenient access to the references available via the internet at the time of publication; however, no assurance is provided that these resources will be accessible in the future.

ABOUTTHEAUTHOR

As a student and young man, Brent Crowder was drawn more to math and science than to social studies, history, and art. Math and science gave him the satisfaction of insight into the qualitative cause-effect relations that describe the physics and chemistry of nature.

As a teen, Brent enjoyed working for his father, who employed him to install thousands of feet of trenching and piping from water wells to and within two fifty-acre fields with sprinkler systems. He also assisted in the operation of the sprinkler systems, as well as the operation and maintenance of construction and farm equipment such as bulldozers, backhoes, plows, and wellhead engines and pumps.

Subsequently, he earned his bachelor of science degree in mechanical engineering and worked professionally during various stages of design, construction, start-up, and operation on coal-fired, geothermal, natural gas, nuclear, and hydroelectric power plants in the United States.

Through his inherent desire to fundamentally understand the mathematics and geometry of nature, he delightfully stumbled upon the number phi, the golden ratio. This number, due to its natural significance, simultaneously created a sense of awe and dismay because such an important natural number was completely absent from his formal education. So, he became what might be described as an inspired "sacred geometry" enthusiast, noting, along with others, that the pyramids of Giza seem linked, internally and externally, all the way to God above, with connectedness to nature and number.

He hopes this book will inspire others as well.

www.ingramcontent.com/pod-product-compliance
Lightning Source LLC
LaVergne TN
LVHW050344210926
848481LV00017B/1446

9798349247293